A DAY UNDERWATER

by Deborah Kovacs

SCHOLASTIC INC.
New York   Toronto   London   Auckland   Sydney

Art direction/design by Diana Hrisinko
Cover art/text art by Walter Wright
Photo layout by Denise Vannucci

No part of this publication may be reproduced in whole or in part,
or stored in a retrieval system, or transmitted in any form or by any means,
electronic, mechanical, photocopying, recording, or otherwise,
without written permission of the publisher.
For information regarding permission, write to
Scholastic Inc., 730 Broadway, New York, NY 10003.

ISBN 0-590-40746-5

Copyright © 1987 by Deborah Kovacs.
All rights reserved. Published by Scholastic Inc.

12   6 7 8 9/9

Printed in the U.S.A.   2 4

First Scholastic printing, May 1987

PHOTO CREDITS

Page 4: Spread, © Robert D. Ballard/Woods Hole Oceanographic Institution; lower, © John Corliss, George Washington University. Page 5: © J. Frederick Grassle/WHOI. Page 8: left, © Rod Catanach/WHOI; right, © C. Dickson/WHOI. Page 9: © Rod Catanach/WHOI. Page 10: © Rod Catanach/WHOI. Page 11: © Robert D. Ballard/WHOI. Page 13: left, © Robert D. Ballard/WHOI; right, © Dudley Foster/WHOI. Page 14: spread, © Kathleen Crane/WHOI; lower, © John Edmond, MIT/WHOI. Page 15: © Dudley Foster/WHOI. Page 16: © Ruth Turner, Harvard/WHOI. Page 17: © WHOI. Page 18: spread, © Fred Speiss/Scripps Institution of Oceanography. Page 19: © J. Childress/University of California, Santa Barbara. Page 20: © Robert D. Ballard/WHOI. Page 21: © WHOI. Page 22: © Roberta Baldwin/SIO. Page 23: upper, © Russ McDuff/WHOI; lower, © Robert Hessler/SIO. Page 24: © WHOI. Page 25: left, © WHOI; right, © WHOI. Page 26: Smith/SIO. Page 27: © J. Frederick Grassle/WHOI. Page 28: left, © John Slade/WHOI; right, © J. Frederick Grassle/WHOI. Page 29: © Rod Catanach/WHOI. Page 30: © John Whitehead/WHOI. Page 31: © John Donnelley/WHOI. Page 32: © Rod Catanach/WHOI.

With special thanks to the Woods Hole Oceanographic Institution, particularly Dr. J. Frederick Grassle, Nancy Green, Shelley Lauzon, Anne Rabushka, and all the Alvin explorers.

Please note that although this book tells the story of one day on Alvin, the photographs are actually from many Alvin dives, from many different parts of the Pacific Ocean.

Where are we? Is this the moon? No! This the bottom of the ocean. We are 12,000 feet underwater.

Everyone knows about the astronauts who explore space. But do you know about the people who explore the ocean floor? They travel in special submarines that are a lot like spaceships.

What is the ocean floor like? What lives there?

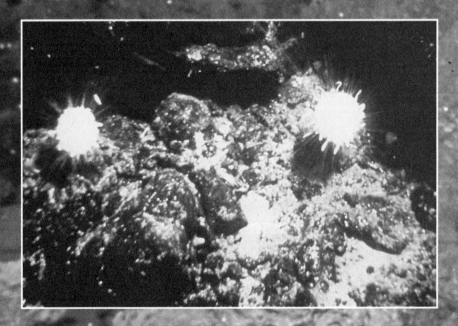

Most of the ocean floor is freezing cold and completely dark. There are strange rocks and even stranger animals. It is like nowhere else on earth.

The ocean floor is so different from where we live that scientists want to learn as much about it as they can. Sometimes they take short trips to the ocean floor in special submarines. Today you will join them.

This is Alvin, the submarine you will ride in. Alvin is one of the deepest-diving boats in the world. It can go down as far as 13,000 feet—more than two miles!

Alvin is more like a spaceship than a boat. It carries tools to help scientists study the world outside its tiny windows. It carries weights that help it sink more quickly. It also carries sample baskets and robot arms to use in experiments. There is hardly any room for the people!

The scientists who will dive with you have spent almost two years getting ready for this day. They have been reading about the ocean floor. They have been studying the work of other scientists. They have worked long hours in their laboratories. At last, it is time to go to sea.

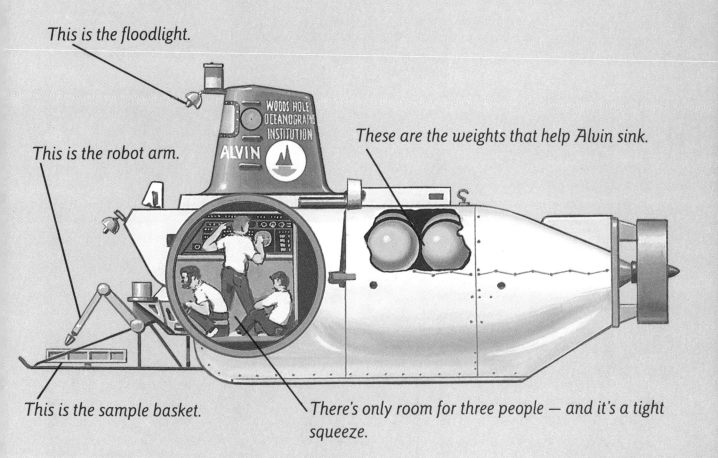

Workers make sure that everything on Alvin is shipshape.

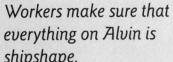

You ride with Alvin and its crew to the middle of the ocean, on a huge ship called the Atlantis II.

Alvin's dive begins very early in the morning. The submarine must return by sunset, or the Atlantis II would have trouble finding it.

Alvin has room for three people — usually two scientists and a pilot. But there is room for you today, too. It is launch time, so climb aboard!

Alvin is lowered over the edge of the Atlantis II, until it rests on the water.

One scientist calls this "the most nervous part." If a big wave came right now, Alvin could be smashed against the Atlantis II.

Down, down, down sinks Alvin — straight down. Light from the surface fades quickly. It will take about two-and-a-half hours to reach the ocean floor. Once there, the scientists will have five hours to do their experiments.

Outside the porthole, a strange fish swims by. It looks like a ghost traveling through the sky on a dark, clear night.

Finally, Alvin arrives on the bottom. You come to a vent, a crack in the crust of the earth. The pilot shines Alvin's bright lights on a group of "black smokers" — underwater chimneys that send up chemicals and very hot water. Inside the smokers it is more than 650 degrees Fahrenheit!

Many kinds of animals live near here. These animals could not live in other parts of the ocean floor. The water would be too cold. But by the vents there is energy and food. Just as we need the warmth and light of the sun, these animals need the heat that comes from below the earth's crust.

Alvin travels to a forest of some of the strangest animals on earth. These are tube worms — 12-feet-long creatures with no eyes, no mouths, and no stomachs, who live in 12-feet-long tubes. They sway in the darkness, soaking up food from the sea water through their thousands of tentacles.

You see an experiment left by scientists on an earlier Alvin dive. It is looking for new animals that have moved to the vent area. The scientists check on the experiment as you pass by.

Alvin uses its underwater light to help the pilot find his way in the utter darkness.

Here is a huge bed of vent clams. These clams are about 10 times bigger than the clams we eat. Their shells are as big as dinner plates! Inside the shells, the clams have bright red bodies.

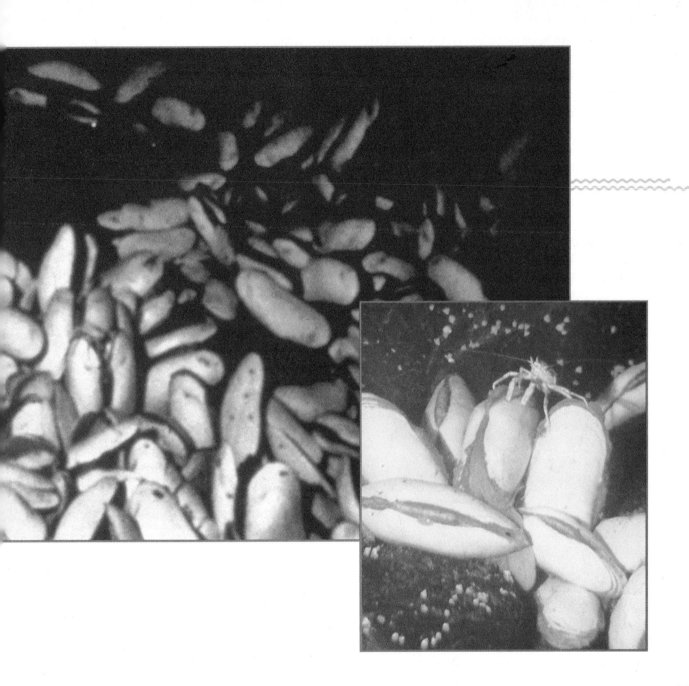

The scientists want to collect some of the clams. Alvin's robot arm will do the collecting. The pilot works a set of switches that control the arm. The arm drops the clams into baskets that hang off the end of Alvin.

Alvin's compass helps the explorers find their way around in the blackness. Without equipment like this, it would be very easy to get lost.

What's that up ahead?

It is a scampering, romping group of white crabs. The crabs have small eyes, but do not need them because their world is always dark. The scientists collect a few crabs in special traps.

Alvin's robot arm uses a "mud grab" — a tool that collects samples of mud from the ocean floor. The arm twists the T-bar handle to pick up the mud. Alvin will take the mud up so scientists can study the tiny animals that live in it.

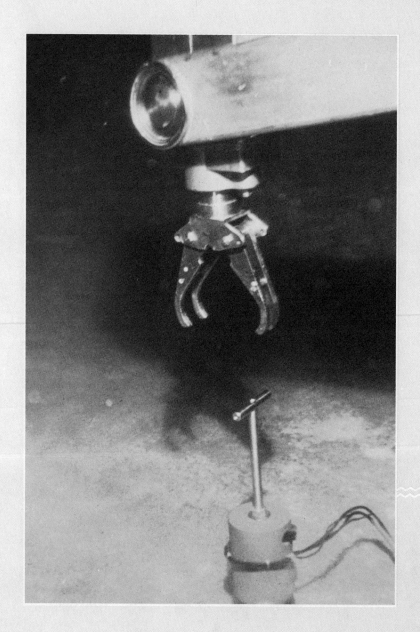

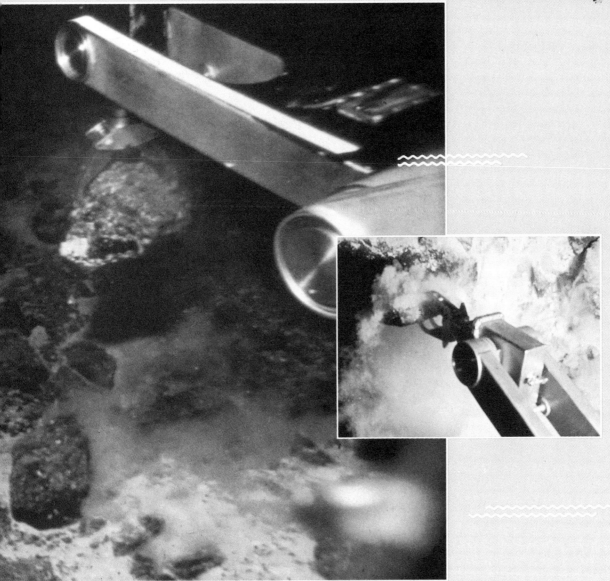

At another spot, the arm collects rocks. The scientists will study the rocks later, to learn more about them.

The animals that live near the vents are very fragile, so scientists work very carefully when they visit the ocean floor.

Here they are measuring the growth rate of mussels. A curious fish swims by.

Sometimes deep-sea scientists worry about changes that people want to make in the ocean.

Some people want to mine the rocks and minerals that are on the ocean floor. Others talk about burying poisonous nuclear wastes there. But scientists know that the animals on the ocean floor are delicate, and would easily be hurt by such big changes.

It is time to head back up. Alvin rises to the surface. There it is met by divers, happy to see that the crew is safe and sound, and excited to hear about the day.

Scientists look at a giant clam and a giant mussel.

A scientist measures the length of a very long tube worm.

Back on the deck of the Atlantis II, the scientists pull the samples from the baskets. Everyone on board crowds around, eager to see what has been found.

The ocean floor is the last place on earth we do not know much about. Though scientists have traveled there for many years, they have still seen only a tiny bit of it.

The work of Alvin scientists is very important. They want to understand the ocean floor and learn how to protect it. With every dive Alvin makes, we learn more and more about this strange world and the beautiful creatures that live there.